Energy
132

在空中捕猎
Fishing in the Air

Gunter Pauli

[比] 冈特·鲍利　著

[哥伦] 凯瑟琳娜·巴赫　绘

贾龙慧子　译

上海远东出版社

丛书编委会

主　任：田成川

副主任：闫世东　林　玉

委　员：李原原　祝真旭　曾红鹰　靳增江　史国鹏

　　　　梁雅丽　孟小红　郑循如　陈　卫　任泽林

　　　　薛　梅　朱智翔　柳志清　冯　缨　齐晓江

　　　　朱习文　毕春萍　彭　勇

特别感谢以下热心人士对童书工作的支持：

匡志强　宋小华　解　东　厉　云　李　婧　庞英元

李　阳　梁婧婧　刘　丹　冯家宝　熊彩虹　罗淑怡

旷　婉　王靖雯　廖清州　王怡然　王　征　邵　杰

陈强林　陈　果　罗　佳　闫　艳　谢　露　张修博

陈梦竹　刘　灿　李　丹　郭　雯　戴　虹

目录

Contents

ZERI Learning Initiative

清晨，一只撒网蛛正在欣赏自己捕获的飞蛾。刚刚过去的这个晚上可是异常繁忙。旁边的一枝郁金香目睹了这只技艺高超的撒网蛛为了养家糊口而进行的这场大捕猎的全过程。

"你昨晚大获全胜啊。恭喜恭喜！"郁金香说。

Early in the morning, a fisherman spider is admiring its catch of moths. It has been a busy night. A tulip has been watching the skilled spider hauling a big catch to feed his family.
"You have been very successful last night. Congratulations!" says the tulip.

你昨晚大获全胜啊。

you have been very successful last night.

……这片田野里有好几百枝郁金香呢。

... there are hundreds of tulips in this field.

"哦，谢谢你！依我看，你一定很擅长养家，因为这片田野里有好几百枝郁金香呢。多漂亮啊！"撒网蛛答道。

"繁衍很多后代和养育后代是有很大区别的。好在我们郁金香非常擅长让黄蜂和蜜蜂帮助我们养育后代。它们帮助我们找到碳元素丰富的土壤，可以供养我们的后代。"

"Oh that is very kind of you. I see that you must be feeding your family very well since there are hundreds of tulips in this field. Beautiful!" responds the spider.

"There is a big difference between having many offspring and babies, and feeding them. We, the tulips, are very good at getting wasps and bees to make babies for us, and find a carbon rich soil to feed us."

"恕我冒昧，请问这些带刺的昆虫具体为你的宝宝们做了些什么呢？"撒网蛛好奇地问道。

"哦，你不知道吗？我会散发一种香味，闻上去就像雌性昆虫最爱的气味。这样，所有的雄性昆虫就会蜂拥而至，在我这里逗留。"

"Excuse me but what do these stinging insects have to do with your babies?" wonders the spider.

"Oh don't you know? I make this perfume that smells like the favourite of the ladies and then all the men come and visit me."

我会散发一种香味……

I make this perfume ...

......它们的身上沾满了我的种子。

... their body is full of my seeds.

"欺骗雄性昆虫让它们相信这里有雌性昆虫，但实际上在你的花瓣上什么也没有？当它们发现被你愚弄了，难道不会很生气吗？"

"我不在乎。它们的刺不会伤害到我。正相反，当它们在我这里搜寻雌性的时候，它们的身上沾满了我的种子。"

"Tricking men to believe that there are ladies when there are none around your petals? Don't they get upset when they realise you fooled them?"

"I don't care. Their stinging energy does not hurt, on the contrary, as they search for the lady, their body is full of my seeds."

"我明白了！所以当下一枝郁金香也散发雌性香味的时候，你们都已得到所需的所有花粉，一个快乐的大家庭即将诞生了。"

　　"你这么快就明白啦。你也是这样对待飞蛾的吗？"

　　"哦不，飞蛾是我的食物。"

"Got it! So when the next tulip also smells like a lady, your flowers get all the pollen needed to create a big and happy family."

"You understand this so quickly. Do you do the same with the moths?"

"Oh no, moths are my meal."

······飞蛾是我的食物。

... moths are my meal.

······我是蜘蛛渔夫吧。

... I am a fisherman spider.

"那你怎么能如此轻易地就捕到它们？"

"你不知道我是蜘蛛渔夫吧。"

"渔夫？那你是怎样在空中捕'鱼'的？"

"So how can you catch them so easily?"

"Don't you know that I am a fisherman spider."

"Fisherman? How do you fish them out of the air?"

"我只是把蛛丝挂出去，飞蛾一旦碰到蛛丝就会被蛛丝粘住。"

"那不可能一晚上就捕到了七次！机会难得，万里挑一，你一定还有诀窍！"

"好吧，我会加一点诱饵。"

"I just hang out a line and the moths passing by stick to it."
"That is impossible seven times in a night! The chance is only one in a million. There must be a trick!"
"Well I am adding some bait."

....把蛛丝挂出去，飞蛾一旦碰到蛛丝就会被蛛丝粘住。

... hang out a line and the moths passing by stick to it.

......飞蛾就会来我这儿寻找伙伴了。

... the moth comes looking for a friend.

"诱饵？你是说你会往空中投放一些飞蛾喜欢的食物，引诱它们向你飞来，于是就撞上了你的蛛丝？"

　　"是的，我和你做的差不多。我在蛛丝上加一点气味，然后飞蛾就会来我这儿寻找伙伴了。"

　　"这么说你悬挂在空中的蛛丝闻上去像是雌性飞蛾？接下来呢？"

"Bait? You mean you throw some food in the air that moths like so they come to you and stick to your line?"

"Well do I do the same as you: I add a little smell to my line, and the moth comes looking for a friend."

"So your line hanging in the air smells like a lady moth? And then what?"

"只要飞蛾一碰到我的蛛丝，它就会立刻被困住。我先用蛛丝把飞蛾卷起来，再用色泽莹润的蛛丝打包带回来，这样我的家人稍后就可以享用了。"

"你是我认识的第一个不需要鱼钩和渔网的渔夫。不，你是我认识的唯一一个在空中捕'鱼'的渔夫。"

……这仅仅是开始！……

"Well as soon as a moth touches my line it is stuck. I reel it in and put a nice silk wrapping around it so the family can have it later".

"You are the first fisherman I know that needs no hooks or nets. No, you are the only fisherman I know that fishes in the air."

… AND IT HAS ONLY JUST BEGUN! …

... AND IT HAS ONLY JUST BEGUN! ...

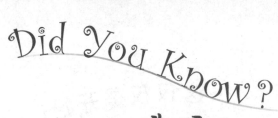

Tulips were introduced to Europe from Turkey in the 16th Century and were considered symbols of nobility, and a proof of wealth.

16世纪，欧洲从土耳其引进郁金香，人们将其视为高贵的象征和财富的证明。

Tulips are members of the lily family and can grow 80 cm tall, taking many colours and shapes. The name tulip is believed to be derived from a Persian word which means turban. During the Ottoman empire it was fashionable to wear tulips on turbans.

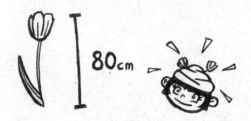

郁金香属于百合科，可长至80厘米高，拥有多种多样的颜色和形状。郁金香这一名称被认为源于波斯语"头巾"。在奥斯曼帝国时期，在头巾上佩戴郁金香是很流行的。

The word tulip in Turkish is "lale", which has the same letters as Allah. That is why the flower became a religious symbol and widely used as a decoration. Tulips have been farmed since the 10th century.

郁金香在土耳其语中用"lale"表示，与真主安拉一词使用相同的字母。这就是郁金香被视为宗教象征，并被广泛用于装饰的原因。自10世纪以来，人们一直在种植郁金香。

The Dutch celebrate the National Tulip Day to mark the beginning of the tulip season. There are an estimated 1.7 billion tulips around the world each year that originated from the Netherlands signalling the arrival of Spring.

荷兰人举办国家郁金香节，以此作为郁金香季节开始的标志。每年春天到来之际，大约有17亿枝郁金香从荷兰运往世界各地。

The fisherman spider looks like a bird dropping and thus remains unnoticed during the day. When females are removed from their nest they produce a very unpleasant smell.

撒网蛛看起来与鸟粪十分相似，因此在白天很难被发现。每当雌性从它们的巢中离开，它们就会散发出一种非常难闻的气味。

The spider has a very good eyesight and will attempt to catch any insect flying by. The spider produces a pheromone for one specific moth.

蜘蛛拥有绝佳的视力，它们试图抓住飞过的任何昆虫。蜘蛛会释放信息素吸引特定的飞蛾。

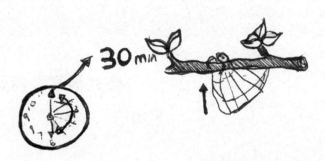

If the hunt was not successful within half an hour, then the spider will eat its "fishing line" and reel out a new one from its abdomen somewhere else.

如果蜘蛛在半小时内没有收获，它们会吃掉自己的"钓线"（蛛丝），然后从腹部的某个地方取出新的。

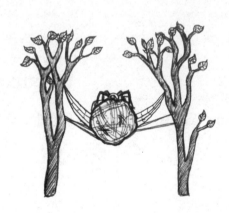

The spider produces bags full with hundreds of eggs. These bags are larger than the mother spider. The male spiders that emerge from the bag are immediately adults.

蜘蛛产卵时会吐丝做成卵囊，每个囊中有数百个卵。有些卵囊甚至比蜘蛛妈妈还大。雄性蜘蛛从卵囊中出来后会迅速成年。

Are you attracted to a particular smell?

你会被某种特殊的气味吸引吗？

How would you feel when you thought you smelled a rose, and instead you find a spider? Would you be upset or disappointed?

当你闻到玫瑰的味道，循着气味却找到了一只蜘蛛，你会有什么感觉？你会生气或者失望吗？

Does it make sense to fish for moths in the air (if you had never heard of the fisherman spider)?

在空中捕猎飞蛾是可能的吗（假定你从未听说过撒网蛛）？

What is the greatest responsibility: bringing children to the world, or ensuring that they have food during their infancy?

哪种行为更有责任感，是把孩子们带到这个世界上，还是确保他们在幼年时期拥有食物？

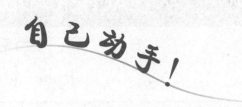

Hang a threat from the ceiling. Put some glue on it. Now make a little paper flying plane and throw the plane towards the sticky threat. How many times have you throw the plane in order to be "caught" by the net? Make an effort to calculate the chance to make the contact. What is the probability and how could you improve your chance to succeed in this challenge. Then you and your team can reflect on the creative approach the fisherman spider needed to deploy in order to secure food for the family.

在天花板上挂一个陷阱，在上面涂抹胶水。再叠一个小的纸飞机，把飞机抛向带有黏性的陷阱。你抛了多少次才让飞机被网"捕住"？试着计算一下取得这个结果的可能性。概率是多少？怎样能提高成功的可能性？此后，你和你的团队可以仔细考虑一只为全家捕食的撒网蛛在部署时可采用的创造性方案。

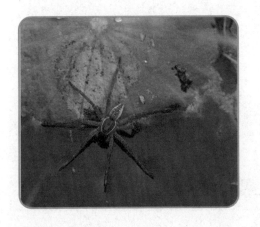

学科知识
Academic Knowledge

生物学	约有160 000种蛾属于鳞翅目，其中大多数都在夜间活动；约有5 000种园蛛科蜘蛛；信息素通过油脂或汗液释放；茉莉花、杨依兰和广藿香的植物油与人类信息素有较强的相似性；几乎所有类型的动物都有信息素，如鱿鱼、龙虾、蚂蚁、鱼、蜥蜴、老鼠；群居昆虫，如蚂蚁、蜜蜂、黄蜂，使用信息素控制群体各个部分的行为。对于拥有超过一百万只蚂蚁或蜜蜂的群体，信息素是女王将消息发送给群体中每一位成员的唯一可行方法；一些捕食者可以伪造信息素来引诱毫无戒心的猎物；郁金香会感染一种被称为马赛克的非致命性病毒，这种病毒不会消灭郁金香的种群，但会使得郁金香花瓣的颜色呈"火焰"状。
化　学	信息素是一种化学物质，会触发社会行为的变化；空气传播的信息素分子会引发同一物种中成员的反应；信息素可建立一种化学通讯方式。
物　理	防守拟态：撒网蛛形似鸟粪，所以在白天很难被注意到；警报信息素普遍存在于昆虫和水生生物之中。
工程学	当蜘蛛的猎物接触到蛛网，对称、高强度的蛛网有助于把撞击的力量均匀地分散开来。
经济学	极少数的信息素被用来控制农业害虫；在害虫（如蛾）对杀虫剂产生抵抗力后，农民开始关注信息素；有些飞蛾的幼虫吃织物（衣服和毯子等），这些织物由羊毛或丝绸等天然纤维制成；有些飞蛾因为它们的经济价值而被养殖，比如家蚕的幼虫——桑蚕。
伦理学	投资者应该了解，比起通过推测赌上未来，停下闻闻花香是更好的选择。
历　史	1959年，诺贝尔奖得主、德国化学家阿道夫·布特南特和他的团队以化学方法鉴定了第一个信息素；17世纪30年代爆发的"郁金香热"是有据可查的市场崩溃之一。
地　理	虽然荷兰使郁金香（尤其是其球茎）在全球范围内商业化，但郁金香原产于非洲北部和土耳其。
数　学	概率是对于事件发生可能性的量度，量化为数字后，它的范围介于不可能（0）和确定（1）之间；概率的计算要使用公理化数学运算方法，即概率论，被用于统计、金融、科学、人工、计算机科学、博弈论中。
生活方式	在家中养花的渴望；花的象征意义。
社会学	在各种各样的传统概念中，花常被认为是美丽、和平、纯洁、生命的象征，也包括回忆；在坟墓前总是能看到花，这是对个人回忆的视觉提醒，也是对死者的祝福。
心理学	人类信息素在增强吸引力的同时也增加了攻击性。
系统论	在一个多生物的免疫系统中，信息素吸引山松甲虫，使它们寄生于树木，树木释放信息素，产生更多的树液，杀死昆虫，以防止它们吃树皮。

情感智慧
Emotional Intelligence

郁金香

郁金香祝贺撒网蛛捕猎成功。郁金香对待行为和责任是有区别的，它有伦理考量，这让它对蜜蜂和黄蜂帮助自己的种族繁殖一事非常满意。郁金香对撒网蛛的无知感到惊讶，但很快它就把原理分享给了撒网蛛。郁金香称赞了撒网蛛的捕猎行为，也很乐于向它学习。郁金香起初不相信在空中捕猎的有效性，并且理性地阐释其概率是非常小的。了解获取食物的独特方式后，对于撒网蛛的能力，郁金香感到非常兴奋并表达出了积极的情绪。

撒网蛛

撒网蛛对郁金香的表扬十分感激，它立即表达了对于郁金香在田野里成功繁育大家庭的称赞。撒网蛛不明白昆虫的作用，但它有提问的自信。撒网蛛坦率地表达了自己的无知，不过它有强烈的好奇心并且持续地询问了更多细节。一旦撒网蛛明白了原理，它就把这套逻辑总结出来。撒网蛛惊讶于郁金香不知道它是怎么获取食物的，同时指出了繁殖和获取食物的区别。撒网蛛以一种简明而务实的态度，花费了很多时间来解释细节，因此赢得了欣赏甚至崇拜。

艺术
The Arts

郁金香是一个很受欢迎的绘画对象。它的吸引力之一就是花瓣上的火焰图案，这是由一种植物病毒引起的。看一看郁金香，现在你能只用铅笔画出黑白的郁金香吗？这些不同种类和形状的火焰在17世纪的郁金香世界里造成了很大程度的狂热，并被认为是这种中东花朵的美丽和独特之处。

思维拓展
Systems: Making the Connections

生存是建立在能获取食物和确保下一代有条件提升生活水平的基础上的。获取食物是至关重要的，不同的物种采用了独特的方法和手段，以求在生态系统中占据一席之地。然而核心工具箱是有限的。自然奇迹的多样性和复杂性最终依赖于一套透明的工具和技术，它们的基础在于已知的物理定律和化学性质的巧妙使用。虽然物理定律没有例外，由此产生的结果是可以预测的，但在化学性质方面，大自然只使用水作为溶剂，在其中与碳相作用。这种简单性和可预测性减少了失败的可能性，打消了对于失败的顾虑，并被运用于特定的环境中，在这里，上千种栖息于相同环境中的生物共同生存着。但也可以将这些蕴含丰富能量的基本工具和材料转化为有着迷人多样性的产品、过程以及结果。正是这种非凡的能力保证了基本的透明和简单，防止演变为戏剧性的多样性。这种多样性使人非常吃惊，同时说明人类社会的进化方式是十分复杂的，可能有数十万种新合成的分子从未被生态系统熟知，经常出现违背（藐视）物理定律的情况。基于这种情况，我们的现代世界对能量的大量需求让人类社会成为产生污染和浪费的源头。因此重要的是，我们不仅要了解大自然，也要不断向大自然学习。

动手能力
Capacity to Implement

让我们做个调查。你使用除臭剂吗？你有没有看过你使用的涂抹在身体上的产品的成分？看一看！是否含有任何金属氧化物，如锌或氧化铝？这些会阻塞你的毛孔。汗液是身体的一种清洁剂，所以排汗非常重要，阻碍这种液体的排出，随着时间的推移会对你的健康有害。来看看来自自然的天然除臭剂吧，这很重要。找出来，列出你喜欢的天然气味，它可以压制住你认为不愉快的气味。分享这些发现，也许继续下去，你会找到你自己的天然除臭剂。

故事灵感来自
This Fable Is Inspired by

托马斯·艾斯纳
Thomas Eisner

托马斯·艾斯纳出生于德国柏林，后移居美国，于哈佛大学获得了生物学学士学位。艾斯纳是动物行为方面的国际权威，被认为是化学生态学之父，他于1957年加入康奈尔大学。他写作（联合写作）了约500篇科学文章、9本关于昆虫研究的书籍，研究昆虫如何交配、捕食猎物并防御捕食者。他的书《热爱昆虫》获得了最佳科学图书大奖，他的电影《秘密武器》被评为最佳科学电影。艾斯纳把昆虫称为化学大师。他让我们更好地了解了蜘蛛织网的过程、投弹甲虫的爆炸性高温喷雾以及它们如何防范捕食者、一些雄性蝴蝶为何分泌某些物质、萤火虫为何不咬人或蜇人。以上这些构成了本册故事以及其他故事的主题。艾斯纳促成了默克公司与哥斯达黎加之间的伙伴关系，旨在为哥斯达黎加生物多样性保护创造收益。他是美国国家科学院院士。

图书在版编目（CIP）数据

冈特生态童书.第四辑：修订版：全36册：汉英对照 /
（比）冈特·鲍利著；（哥伦）凯瑟琳娜·巴赫绘；
何家振等译.—上海：上海远东出版社，2023
书名原文：Gunter's Fables
ISBN 978-7-5476-1931-5

Ⅰ.①冈… Ⅱ.①冈… ②凯… ③何… Ⅲ.①生态环
境−环境保护−儿童读物—汉、英 Ⅳ.①X171.1-49

中国国家版本馆CIP数据核字（2023）第120983号
著作权合同登记号图字09-2023-0612号

策　　划　张　蓉
责任编辑　曹　茜
封面设计　魏　来 李　廉

冈特生态童书
在空中捕猎
[比]冈特·鲍利　著
[哥伦]凯瑟琳娜·巴赫　绘

贾龙慧子　译

记得要和身边的小朋友分享环保知识哦！
八喜冰淇淋祝你成为环保小使者！